W9-BLD-950

WE CAN READ about NATURE!™

CAN YOU FIND US?

by ANITA HOLMES

BENCHMARK BOOKS

MARSHALL CAVENDISH
NEW YORK

With thanks to
Susan Jefferson, first grade teacher at Miamitown Elementary, Ohio, for sharing her innovative teaching techniques in the Fun with Phonics section.

Benchmark Books
Marshall Cavendish Corporation
99 White Plains Road
Tarrytown, New York 10591

Photo research by Candlepants Inc.

Cover photo: *The National Audubon Society Collection / Photo Researchers, Inc.*: E. R. Degginger

The photographs in this book are used by permission and through the courtesy of: *The National Audubon Society Collection / Photo Researchers, Inc.:* Michael Lustbader, 4; Gilbert S. Grant, 5; Scott Camazine, 6; J. H. Robinson, 7; Tom McHugh, 8, 15(top); C. K. Lorenz, 9; Nuridsany & Perennou, 10; Fred McConnaughy, 11(top); E. R. Degginger, 11(bottom), 12; Fletcher & Baylis, 14; Dan Guravich, 15(bottom); Leonard Lee Rue III, 16-17; Tim Davis, 18; Stephen J. Krasemann, 20; Ken M. Highfill, 21(top); Dr. Paul Zahl, 21(bottom); Art Wolf, 22, 23; Cosmos Blank, 24; Kent & Donna Dannen, 27; Suzanne & Joseph Collins, 28.

Library of Congress Cataloging-in-Publication Data

Holmes, Anita, date
Can you find us? / by Anita Holmes.
p. cm.– (We can read about nature!)
Summary: Presents a simple discussion of how nature uses color and shape to protect animals from predators with examples such as the chameleon, tree frog, and Arctic fox.
ISBN 0-7614-1108-9
1. Camouflage (Biology)–Juvenile literature. [1.Camouflage (Biology)] I.Title.
QL759 .H59 2001 591.47'2–dc21 99-057014

Printed in Italy

1 3 5 6 4 2

Look for us inside this book.

chameleon
dragonfly
fox
frog
gecko
grasshopper
katydid
spider
stonefish
treehopper
walkingstick
zebra

Who's hiding there?
We are.
We are insects.

A green dragonfly, newly born

A grasshopper nymph

Our color blends with
the green of the leaves.
This makes us hard
to see.

I'm a tiny tree frog.
I eat insects.
Snakes eat frogs.

A green tree frog in the Everglades

Why am I hiding?
So a snake won't
find *me*.

Can you find us?
We are shades of brown.

A horned frog in Malaysia

A foam nest tree frog in South Africa

Our color matches the place
where we make our home.
This is called camouflage.

Can you find us . . .

A crab spider

on the flower . . .

on the coral . . .

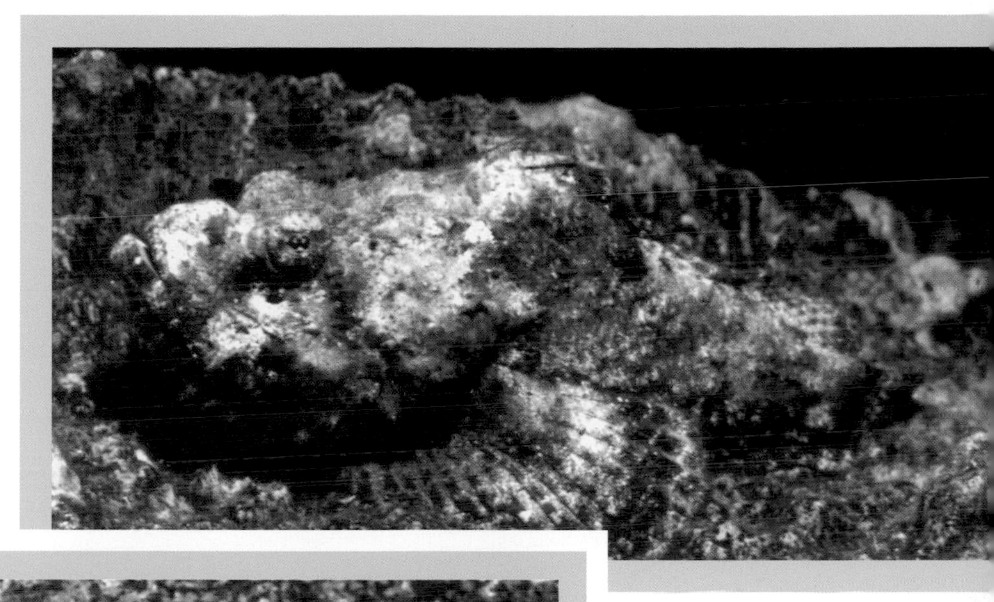

A false stonefish

on the rock?

A Brook's gecko

I'm a chameleon, a kind of lizard.

Jackson's chameleons in West Africa

I can turn dark
when it's hot.
Or green when
it's cool.
I'm brown and
black when
I'm a baby.

I live on the tundra.
Summers are short,
and winters are long and snowy.
I change my coat . . .

Arctic fox cubs

in summer . . .

fall . . .

and
winter.

Shhh!
My mother told me to keep still.

A white-tailed deer fawn

My spots are my disguise.
When I'm grown, my
spots will disappear.

We are zebras.
We live in a herd.
We know each other by our stripes.

If a hyena or lion
comes near, we stand
close together.
Will our stripes
confuse them?

Can you find us?

A dead-leaf katydid

I'm a leaf!

I'm a twig!

A treehopper

A walkingstick

I'm a thorn!

Wherever you go—

Impalas

to Africa . . .

Walkingstick

to South America . . .

A boa constrictor

all the way to the Arctic Circle—

A white-tailed ptarmigan

—we are there!

A green tree frog

We are hiding everywhere.

fun with phonics

How do we become fluent readers? We interpret, or decode, the written word. Knowledge of phonics—the rules and patterns for pronouncing letters—is essential. When we come upon a word we cannot figure out by any other strategy, we need to sound out that word.

Here are some very effective tools to help early readers along their way. Use the "add-on" technique to sound out unknown words. Simply add one sound at a time, always pronouncing previous sounds. For instance, to sound out the word **cat**, first say **c**, then **c-a**, then **c-a-t**, and finally the entire word **cat**. Reading "chunks" of letters is another important skill. These are patterns of two or more letters that make one sound.

Words from this book appear below. The markings are clues to help children master phonics rules and patterns. All consonant sounds are circled. Single vowels are either long –, short ˘, or silent /. Have fun with phonics, and a fluent reader will emerge.

Short "o" words:

f r ŏ g   h ŏ t   r ŏ ck   s p ŏ t s

When two vowels are together, the first one does the talking, and the second one goes walking. Talking vowels are long vowels that say their name. Walking vowels do not say anything; they are silent.

c ō a̸ t   ē a̸ t   g r ē e̸ n

k ē e̸ p   l ē a̸ f   l ē a̸ v e̸ s

s ē e   t r ē e   w ā y

Bossy "ar" says the letter name R.

h a r d (ar = R)   d a r k (ar = R)

A magic "e" at the end of the word stays silent, but makes the vowel that comes before it long. (All long vowels say their names.)

c l ō s e   h ō m e

m ā k e   s n ā k e

fun facts

- Lions prey on zebras. They see in black and white. This is why zebras' black and white stripes confuse them.
- To stay safe, some butterflies look like the poisonous butterflies that birds will not eat.
- Decorator crabs hide from prey by putting all sorts of plants and animals on their backs—kelp, sponges, seaweed, anemones. They even grow algae on their shells.
- Soldiers sometimes wear camouflage clothes. The idea came from animals.

glossary/index

Africa	the land south of Europe 22
Arctic Circle	near the North Pole 26
blend	mix together 5
camouflage	blend with background 9
coat	animal fur 14
confuse	mix up 19
disappear	pass from view 17
disguise	change or hide 17
herd	a group of animals 18
hide	be out of sight 4-5, 7, 29
lizard	a kind of reptile 12
shade	a little different color 8
South America	the land south of North America 24
tundra	flat, frozen land 14

about the author

Anita Holmes is both a writer and an editor with a long career in children's and educational publishing. She has a special interest in nature, gardening, and the environment and has written numerous articles and books for children on these subjects. A number of her books have won commendations from the *American Library Association*, the *National Science Teachers Association*, and *The New York Public Library.* She lives in Norfolk, Connecticut.